Beginning Algebra

Real Numbers, Algebraic Expressions, Linear Equations & Graphs

ALSO BY WILLIAM R. PARKS

Computer Mathematics

1 + 1 = 10 Introduction to Computer Number Bases

Introduction to Sets and Flowcharts

Introduction to Logic

1 + 1 = 1 An Introduction to Boolean Algebra and Switching Circuits

Statistics

Introduction to Statistics

Introduction to Gambling Theory

Beginning Algebra

Real Numbers, Algebraic Expressions, Linear Equations & Graphs

William R. Parks, B.S., M.S., Ed.M.

Hershey Books

www.wrparks.com

Cover

The four operations of arithmetic are displayed on the cover: addition, subtraction, multiplication and division. These four basic operations of arithmetic are also used in algebra.

To: Bobby, Brittmi, Grant, Isabella, Joey, John, Jordyn, Mason, Preston, Steve & Stephanie

ISBN: 978-0-88493-030-3

Library of Congress Control Number: 2016901377

Introduction

Beginning Algebra was originally published by Williamsville Publishing Company as part of their popular audio-visual *Tape 'n Text Math Series.* Material in the series was submitted for review in *The Mathematics Teacher* and received praise: "The development was well done." .

This paperback is intended for classroom teachers, students and as a reference book for libraries and learning centers.

Elementary algebra includes the study of basic operations found in arithmetic such as addition, subtraction, multiplication and division. What's new is that letters of the alphabet (called variables) are introduced to stand for numbers as used in business formulas to calculate interest.

The letters, called variables stand for numbers such as integers 1, 2, 3 or decimal numbers such as .5, 1.5, 2.7 etc. The basic rules of arithmetic apply in elementary algebra. However, new concepts are added to arithmetic such as: reasoning about relationships, generalizations about these relationships and the application of logical thinking.

Sometimes math textbooks present an overwhelming amount of information on a given topic. In this book we have simplified explanations and also give examples that are easier to comprehend. Also, new information is presented in short sections with immediate testing.

This form of personalized instruction is often used in on-line Internet based courses for distance learning. A small amount of information is presented in each section before advancing to the next section. Exercises are listed after several sections followed by the answer key.

PROPERTIES OF THE REAL NUMBER SYSTEM

1. The set of real numbers contains several important <u>subsets</u>. Given below is a list of some of the important subsets.

   (a) $\{1, 2, 3, 4, 5, 6, 7, \ldots\}$ is the set of <u>counting numbers</u>.

   (b) $\{-1, -2, -3, -4, -5, \ldots\}$ is the set of <u>negative integers</u>.

   (c) $\{\ldots, -3, -2, -1, 0, 1, 2, 3, \ldots\}$ is the set of <u>integers</u>.

   (d) $\{$numbers that can be expressed as a ratio of two integers$\}$ is the set of <u>rational numbers</u>. e.g. $\frac{1}{2}, \frac{-2}{3}, \frac{5}{4}, \frac{10}{2}$.

   (e) $\{$numbers which cannot be expressed as a ratio of two integers$\}$ is the set of <u>irrational numbers</u>. e.g. $\sqrt{2}, \sqrt{3}, \sqrt{5}, \sqrt{10}$, and others whose roots are not perfect.

   NOTE: The <u>union</u> of the above sets is the set of <u>real numbers</u>.

   – – – – – – – – – – – –

2. The set of real numbers is an <u>infinite set</u>. It contains an inexhaustible number of elements.

   – – – – – – – – – – – –

3. The system of real numbers is a <u>field</u>. In mathematics the term "field" is used for any set of numbers with two operations, addition and multiplication, that satisfy certain basic laws which are called <u>field axioms</u>.

   – – – – – – – – – – – –

4. In the following frames let the letters a, b, and c stand for real numbers. For example, if a and b are real numbers, then their sum, a + b also represents a real number.

   – – – – – – – – – – – –

5. <u>Closure</u>. The set of real numbers is <u>closed</u> under addition, subtraction, multiplication and division (excluding division by zero).

EXAMPLES:

(a) The set of real numbers is <u>closed</u> under addition.
That is, the sum a + b of two real numbers is always
a real number. (This is the <u>Closure Law of Addition.</u>)
If the sum of two numbers is a number we can find in the same
set, then the set is closed during addition.

(b) The set of odd numbers $\{1, 3, 5, 7, 9, 11, \ldots\}$ is <u>not</u>
closed under addition. e.g. $3 + 5 = 8$ (8 is not odd)

(c) The set of rational numbers is closed under addition.

$$\frac{2}{3} + \frac{4}{5} = \frac{(2 \cdot 5) + (3 \cdot 4)}{(3 \cdot 5)} = \frac{22}{15} ; \quad \frac{a}{b} + \frac{c}{d} = \frac{ad + bc}{bd}$$

(d) The rational numbers are also closed under multiplication.

$$\frac{2}{3} \cdot \frac{4}{5} = \frac{8}{15} \quad \text{In general,} \quad \frac{a}{b} \cdot \frac{c}{d} = \frac{ac}{bd}$$

(e) The set of counting numbers $\{1, 2, 3, 4, \ldots\}$ is <u>not</u>
closed under subtraction. e.g. $3 - 5 = -2$
-2 is not a counting number, therefore the set of counting numbers
is not closed under subtraction.

- - - - - - - - - - - - - - -

6. The <u>Associative</u> Laws. If a, b, c are real numbers, then

$(a + b) + c = a + (b + c)$ Associative law means it does not

$(a \cdot b) \cdot c = a \cdot (b \cdot c)$ matter how we group numbers.
The results are the same.

EXAMPLES:

(a) $(6 + 4) + 3 = 6 + (4 + 3)$
$\qquad 10 \quad + 3 = 6 + \quad 7$
$\qquad\qquad 13 \quad = \quad 13$

(b) $(2 \cdot 3) \cdot 4 = 2 \cdot (3 \cdot 4)$
$\qquad 6 \quad \cdot 4 = 2 \cdot 12$
$\qquad\qquad 24 \quad = \quad 24$

(c) The set of real numbers is <u>not</u> associative with respect
to subtraction and division.

$(6 - 4) - 3 \neq 6 - (4 - 3)$
$\qquad 2 - 3 \neq 6 - 1$
$\qquad\quad - 3 \neq 5$

$(8 \div 4) \div 2 \neq 8 \div (4 \div 2)$
$\qquad 2 \div 2 \neq 8 \div 2$
$\qquad\qquad 1 \neq 4$

7. The Commutative Laws. If a, b are real numbers, then

$a + b = b + a$

$a \cdot b = b \cdot a$

EXAMPLES:

(a) $3 \cdot 4 = 4 \cdot 3$ Interchanging numbers does not change their product. Multiplication is commutative.

(b) $4 + 6 = 6 + 4$ Addition is commutative.

(c) $5 - 2 \neq 2 - 5$

$\quad 3 \neq -3$ Subtraction is not commutative.

(d) $8 \div 4 \neq 4 \div 8$

$\quad 2 \neq \dfrac{4}{8}$ Division is not commutative.

---

8. The Identity Laws. For every real number a ,

$a + 0 = 0 + a = a$ e.g. $0 + 4 = 4$

$a \cdot 1 = 1 \cdot a = a$ e.g. $1 \cdot 6 = 6$

Zero is called the additive identity.
One is called the multiplicative identity.

---

9. The Inverse Laws. For every real number a,

$a + (-a) = 0$ e.g. $5 + (-5) = 0$

$a \cdot \dfrac{1}{a} = 1$ e.g. $5 \cdot \dfrac{1}{5} = 1$ $\quad (a \neq 0)$

(-5) is called the additive inverse of 5 or the negative of 5.
$\dfrac{1}{5}$ is called the multiplicative inverse of 5 or the reciprocal of 5.

---

10. The Distributive Law. If a, b, c, are real numbers, then

$a \cdot (b + c) = (a \cdot b) + (a \cdot c)$

EXAMPLES:
(a) $3(a + b) = 3a + 3b$
(b) $4(2a + b) = 8a + 4b$
(c) $6(y + 1) = 6y + 6$
(d) $7(x + 2y) = 7x + 14y$
(e) $\frac{1}{2}(8x + 4) = \frac{8x}{2} + \frac{4}{2} = \frac{^4\cancel{8}x}{\cancel{2}} + \frac{^2\cancel{4}}{\cancel{2}} = 4x + 2$

– – – – – – – – – – –

11. Some other basic properties of the real number system are as follows:

(1) For every real number a, $a \cdot 0 = 0$.

EXAMPLES:
(a) $4 \cdot 0 = 0$
(b) $0 \cdot (-7) = 0$ (Zero is a <u>multiplicative annihilator.</u>)

(2) For every real number a, $(-1) \cdot a = -a$.

EXAMPLES:
(a) $(-1) \cdot 3 = -3$
(b) $(-1) \cdot (-4) = -(-4) = 4$

(3) If a, b are positive real numbers, then $a \cdot (-b) = -ab$ and $(-a) \cdot (-b) = ab$.

EXAMPLES:
(a) $3 \cdot (-4) = -12$
(b) $(-3) \cdot (-4) = 12$

(4) If a, b are real numbers, then $a - b = a + (-b)$.

EXAMPLES:
(a) $4 - 3 = 4 + (-3) = 1$
(b) $(-4) - 3 = (-4) + (-3) = -7$
(c) $3 - 4 = 3 + (-4) = -1$
(d) $3 - (-4) = 3 + [-(-4)] = 3 + 4 = 7$

– – – – – – – – – – –

SUMMARY OF PROPERTIES OF REAL NUMBERS

Property	Addition	Multiplication
associative	$(a+b) + c = a + (b+c)$	$(ab)c = a(bc)$
commutative	$a + b = b + a$	$a \cdot b = b \cdot a$
identity	$a + 0 = a$	$a \cdot 1 = a$
inverse	$a + (-a) = 0$	$a \cdot 1/a = 1$
distributive	$a (b+c) = ab + ac$	$(a+b) c = ac + bc$

closure: The set of real numbers is closed under addition and under multiplication. The sum or product of two real numbers equals a real number. The set of odd numbers is not closed under addition. If you add two odd numbers you can get an even number.

EXERCISE 1: Properties of the Real Number System

Answer all questions before checking the answer key on the next page.

1. Which of the following are rational numbers?
(a) 2 (b) – 7 (c) 2/3 (d) .25 (e) $\sqrt{3}$ (f) All of them.

2. Which of the numbers in question 1 are real numbers?

Questions 3 to 11 – True or False:

3. The set of even numbers is closed under addition.

4. The set of odd numbers is closed under addition.

5. The set of positive integers is not associative with respect to subtraction.

6. Subtraction and division are not commutative.

7. The identity element for addition is zero.

8. The identity element for multiplication is one.

9. The product of two negative numbers is always positive.

10. A negative number times a positive number is negative.

11. Any number times zero is zero.

12. What is the additive inverse of each number.
(a) -8 (b) 10 (c) ½ (d) 2.5 or 2 ½ or 5/2 (e) – .75 or 3/4

13. What is the mutliplicative inverse of each number in question 12?

14. Use the distributive property in the following expressions:
(a) 2(a + b) (b) -3(3x + 5y) (c) 1/3 (9x + 6y)

ANSWERS FOR EXERCISE 1:

1. (a), (b), (c) and (d)

2. (a), (b), (c), (d), and (e)

3. True. The sum of two even number is an even number.

4. False. The sum of two odd number is an even number.

5. True. They do not produce the same values.

6. True. Subtraction and division are not commutative.

7. True. $a + 0 = a$

8. True. $a \cdot 1 = a$

9. True. The product of two negative numbers is always positive.

10. True.

11. True

12. (a) +8 (b) – 10 (c) – ½ (d) – 2.5 or – 2 ½ or – 5/2
    (e) +.75

13. (a) – 1/8 (b) 1/10 (c) 2 (d) 2/5 (e) 4/3

14. (a) 2a + 2b (b) -9x – 15y (c) 3x + 2y)

ALGEBRAIC EXPRESSIONS

1. An algebraic expression is a <u>symbol</u> or a <u>finite collection</u> of symbols.

   EXAMPLES OF SYMBOLS FOUND
   IN ALGEBRAIC EXPRESSIONS:

   (a) numbers: $3, -7, \frac{1}{2}, \sqrt{3}$

   (b) variables: x, y, z, a, b
   (c) operations: $+, -, \cdot, \div, \sqrt{}$
   (d) aggregation symbols: parantheses ( ), brackets [ ], braces { }

   - - - - - - - - - - - -

2. We <u>combine</u> the symbols mentioned in Frame 1 to form algebraic expressions.

   EXAMPLES:
   (a) $x + 2y$
   (b) $\dfrac{5a - 3b}{7 + 4x}$
   (c) $3x^2 - 12$
   (d) $\sqrt{2m^2 + 4n}$
   (e) $(2x - 3y) \cdot (5x + 7y)$

   - - - - - - - - - - - -

3. All individual variables and numbers in algebraic expressions along with the sign (+ or -) are called <u>terms</u>.

   EXAMPLES:
   (a) $2y$ is a term. (A "+" sign is assumed to precede it.)
   (b) $-3xy$ is a term.
   (c) The expression $5xy + 7y$ has two terms: "+5xy" and "+7y."
   (d) $3x^2y$ is a term.
   (e) $3x^2 + 4x - 12$ has three terms.

   - - - - - - - - - - - -

4. Algebraic expressions are classified by name according to the number of terms that are present.

EXAMPLES:
(a) 10a, 3xy, and y are called <u>monomials</u>.
(b) 2x - 7y and 5a + 2b are <u>binomial</u> expressions.
(c) x^2 - 4x + 4 is an example of a <u>trinomial</u>.
(d) A general name for an expression with two or more terms is <u>multinomial</u>.

— — — — — — — — — — — — — —

5. A variable is a letter symbol that stands for <u>numbers</u>. These numbers are found in a well-defined set called the <u>replacement set</u>.

EXAMPLES:
(a) Consider the algebraic expression "2m + 1." If the <u>replacement set</u> for m = { 1, 5, 7 }, then the possible values of 2m + 1 are as follows:
2m + 1 = 3 when m = 1
2m + 1 = 11 when m = 5
2m + 1 = 15 when m = 7
(b) Let the replacement set for n = { positive numbers } in the algebraic expression $2n + n^3$. The value of this expression will always be positive.
(c) Let the replacement set for n = { negative numbers } in the expression $2n + n^3$. The value of this expression will always be negative.

— — — — — — — — — — — — — —

6. A <u>constant</u> is a symbol whose replacement set consists of only one number.

EXAMPLES:
(a) All numbers in algebraic expressions are constants.
(b) Letters which are given only one value in an algebraic expression are considered to be constants.
(c) In the expression "$2 \pi r$, 2 and π are constants.

EXERCISE 2: Algebraic Expressions

Answer all questions before checking the answer key.

1. Give three examples of each of the following symbols found in algebraic expressions:
   (a) numbers: _____, _____, _____
   (b) variables: _____, _____, _____
   (c) operations: _____, _____, _____
   (d) aggregation symbols: _____, _____, _____

2. How many times does multiplication occur in the following expression? $(3x^2 - 2y) \cdot (m^3 + 4xy)$

3. Classify the following algebraic expressions by using one or more of the following names: monomial, binomial, trinomial and multinomial.

   (a) $3x^2$ _____

   (b) y _____

   (c) $\dfrac{3m}{n} + 4mn$ _____

   (d) $x^2 - 5x - 50$ _____

   (e) $5a + 2b - 3c + 2d^2$ _____

4. In the expression "n + 1" suppose that the replacement set for n is the set of even counting numbers: $\{2, 4, 6, 8, ..\}$. How does this replacement set affect the values of the expression n + 1? _____
   _____

5. Find the value of each of the following expressions if the replacement set for each variable is: $\{2, 5, 10, 15\}$.

   (a) x^2 (b) x^3 (c) $3y + 10$ (d) $2b - 20$

6. A constant is a symbol whose replacement set consists of only one number. True or false?

ANSWERS FOR EXERCISE 2:

1. (a) 3, $\frac{1}{2}$, -10 (any real numbers)

    (b) a, b, c (any letter)
    (c) +, -, •, ÷, $\sqrt{\ }$
    (d) parantheses ( ), brackets [ ], braces { }

2. eight times

3. (a) monomial
    (b) monomial
    (c) binomial or multinomial
    (d) trinomial or multinomial
    (e) multinomial

4. The values are odd counting numbers, e.g. when n = 2 then "n + 1" takes on a value of "3"; when n = 4 then n + 1 = 5.

5. (a) 4, 25, 100, 225
    (b) 8, 125, 1000, 3375
    (c) 16, 25, 40, 55
    (d) -16, -10, 0, +10

6. True

7. A <u>factor</u> is a quantity (number or variable) which is multiplied by another quantity to form a product.

EXAMPLES:

(a) In the expression "5ab" there are three factors: "5", "a", and "b."

(b) 6 and 2 are factors of 12.

(c) In $5y^3$ there are four factors. $5 \cdot y \cdot y \cdot y = 5y^3$

(d) $(x + 2y)(x - 3y)$ This represents two factors. Each factor is a binomial.

(e) A monomial factor and a binomial factor may also be combined: $3y^2(2x + 7y)$.

- - - - - - - - - - - - -

8. Verbal expressions can be translated into algebraic expressions. Let variables stand for the unknown and let numbers stand for the constants.

EXAMPLES:

(a) If John is some unknown age, then how old will he be 5 years from now?
Let x represent John's present age, then his age 5 years from now will be: "x + 5."

(b) "Twice his present age" would be "2x."

(c) Five years ago how old was John? (ANSWER: x - 5)

(d) What is half John's age? (ANSWER: $\frac{1}{2}x$)

(e) How old will John be y years from now? (ANSWER: x + y)

(f) Mary and Michael have a joint saving account of $500. Michael's portion is $d. What is Mary's portion?
(ANSWER: $500 - $d or $(500 - d))

(g) If Carol earns $d per month, then what will she earn in a year? (ANSWER: 12 · $d or simply 12d)

(h) If sugar costs $d per pound, then the cost of 10 pounds is 10d. For $5 you can buy $\frac{5}{d}$ pounds.

If the price is reduced $1 per pound and you purchase five pounds, the cost in dollars is represented by 5(d - 1).

EXERCISE 3: Algebraic Expressions

Answer all questions before checking the answer key.

1. Name all factors in the expression $3x^2y$. _____

2. Name as many <u>pairs</u> of numbers as you can think of which are factors of 24. _____

3. Name a three factor combination of 24. _____

4. Name four factors of 24. _____

5. In the expression $2x^2(3x - 7)$, the monomial factor is _____ and the binomial factor is _____ .

6. If Joe and Jenny earn together $2,000 per month and Joe's salary is $d per month, then write each verbal expression given below as an algebraic expression:
   (a) the amount Jenny earns per month _____
   (b) the amount Jenny earns per year _____
   (c) the amount Jenny earns in two years _____
   (d) Jenny's salary increases $100 per month _____
   (e) the amount Joe earns per year _____
   (f) Joe's salary decreases $100 per month _____
   (g) the amount Joe earns per year if his salary increases $200 per month _____

ANSWERS FOR EXERCISE 3

1. $3 \cdot x \cdot x \cdot y$ (four factors)

2. $2 \cdot 12$; $3 \cdot 8$; $1 \cdot 24$; $4 \cdot 6$

3. $2 \cdot 2 \cdot 6$; $2 \cdot 3 \cdot 4$

4. $2 \cdot 2 \cdot 2 \cdot 3$

5. $2x^2$, $(3x - 7)$

6. (a) \$2000 - \$d or simply 2000 - d
   (b) $12(2000 - d)$
   (c) $24(2000 - d)$
   (d) $(2000 - d) + 100$ or $2100 - d$
   (e) 12d
   (f) d - 100
   (g) $12(d + 200)$

9. An algebraic expression can be translated to English by knowing the meaning of mathematical symbols. The process is the reverse where we went from English to Algebra.

Here we go from Algebra to English.

EXAMPLES:

(a) $2n - 5$ means "five less than two-times-some-number."

(b) $3 + x^2$ means "three increased by the square of a number."

(c) $\frac{1}{2}(x - 6)$ means "one-half the difference between a number and six."

(d) $2x + 5y$ means "two-times-a-number increased by five-times-a-number."

(e) $\frac{n}{3}$ or $n \div 3$ means "a number divided by three."

- - - - - - - - - - - - -

10. A <u>polynomial expression</u> or simply a <u>polynomial</u> is an algebraic expression which restricts itself to the operations of addition, subtraction, and multiplication. A polynomial is the sum of a finite number of terms. The terms are restricted to be the product of numbers and variables.

EXAMPLES:

(a) $x^2 + 7x - 30$ is a polynomial since it involves only the operations of addition, subtraction, and multiplication.

(b) $x^2 + \frac{7}{x} - 30$ is <u>not</u> a polynomial because the second

term $\frac{"7"}{x}$ involves division by a variable.

(c) $x^2 + 7\sqrt{x} - 30$ is <u>not</u> a polynomial because the second term includes a variable under a square root sign.

(d) $\frac{1}{2}x^2 + \sqrt{7}\ x - 30$ is a polynomial, even though the division and root sign appear, since $\frac{1}{2}$ and $\sqrt{7}$ are numbers.

EXERCISE 4: Algebraic Expressions

Answer all questions before checking the answer key

1. Translate the following algebraic expressions into English:

   (a) $3x^2$ _____

   (b) $2n + 4$ _____

   (c) $\frac{1}{4}(x - 4)$ _____

   (d) $\frac{2n}{3}$ _____

   (e) $5a + 2b$ _____

   (f) $x^2 + (7x - 15)$ _____

2. Determine which of the following are polynomial expressions:

   (a) $\frac{1}{4}x^2 + 3x - 7$

   (b) $x^3 - \sqrt{4}\, x^2 + 5x - 17$

   (c) $\sqrt{x} + 2x^2 - 7$

   (d) $\dfrac{5}{3x^2 - 10}$

   (e) $x^3 + 2x^2 - 5x + 4$

ANSWERS FOR EXERCISE 4:

1. (a) three times the square of a number $3 \cdot x \cdot x$
   (b) two-times-a-number increased by four $2 \cdot n + 4$
   (c) one-fourth the difference between a number and four
   (d) two-times-a-number divided by three $(2 \cdot n) / 3$
   (e) five-times-a-number increased by two-times-a-number
   (f) the square of a number increased by the difference between seven-times-a-number and fifteen

2. The polynomial expressions are: (a), (b), and (e).

LINEAR EQUATIONS AND GRAPHS

1. Compare these two equations!

    $x + 3 = 7$ "x plus three equals seven"

    $x^2 + 3 = 7$ "x squared plus three equals seven"

    The exponent of "x" in the first equation is assumed to be one.

    – – – – – – – – – – – –

2. $x + 3 = 7$ is called a <u>first</u> degree equation.

    We will be discussing this type of first degree equation. In the above equation, the exponent of the variable "x" is understood to be one. This type of equation is also called a linear equation because, it will graph as a straight line.

    $x^2 + 3 = 7$ is called a <u>second</u> degree equation.

    In this equation, the exponent of the variable "x" is two. This type of equation will graph as a curved line.

    – – – – – – – – – – – –

3. Let's compare the following two first degree equations. The coefficient of x in the first equation is understood to be one. The coefficient of x in the second equation is 2.

    $x + 3 = 7$ Some number "x" increased by 3 equals 7.

    $2x + 3 = 7$ Two times some number increased by 3 equals 7.

    – – – – – – – – – – – –

4. To find the value of the variable x in the first equation subtract 3 from each side of the equation. The solution is "x equals 4."

    $$\begin{aligned} x + 3 &= 7 \\ -3 &= -3 \quad \text{subtract} \\ \hline x + 0 &= 4 \\ x &= 4 \quad \underline{\text{solution}} \end{aligned}$$

    – – – – – – – – – – – –

5. To verify this solution, substitute 4 for "x" in the original equation.

    $x + 3 = 7$

    $(4) + 3 = 7$

    $7 = 7$ Four is the correct answer because $7 = 7$.

6. To solve the equation $x - 3 = 7$, add 3 to both sides.

$$
\begin{array}{rl}
x - 3 = & 7 \\
+\ 3 = & +3 \quad \text{add} \\
\hline
x + 0 = & 10 \\
x = & 10 \quad \underline{\text{solution}}
\end{array}
$$

- - - - - - - - - - - - -

7. To verify that 10 is the right answer, subsitute 10 for "x."

$$
\begin{array}{l}
x \ - 3 = 7 \\
(10) - 3 = 7 \\
\qquad\quad 7 = 7 \quad \underline{\text{check}} \quad \text{Therefore } x = 10 \text{ is a correct solution.}
\end{array}
$$

- - - - - - - - - - - - -

8. Compare the two methods we used to solve the two equations.
In the equation on the left we subtraced 3 and in the equation on the
right we added 3 to obtain the right answer.

$$
\text{COMPARE:}\quad
\begin{array}{rl}
x + 3 = & 7 \\
-\ 3 = & -3 \quad \text{subtract} \\
\hline
x + 0 = & 4 \\
x = & 4
\end{array}
\qquad
\begin{array}{rl}
x - 3 = & 7 \\
+\ 3 = & +3 \quad \text{add} \\
\hline
x + 0 = & 10 \\
x = & 10
\end{array}
$$

- - - - - - - - - - - - -

9. To solve the equation $2x + 3 = 7$, see below steps. We first subtract
3 from each side of the equation and are left with $2x = 4$. Next, we
divide each side of the equation by 2. The bottom 2 cancels out, and
we obtain the solution which is $x = 2$.

$$
\begin{array}{rl}
2x + 3 = & 7 \\
-\ 3 = & -3 \quad \text{subtract} \\
\hline
2x + 0 = & 4 \\
2x = & 4 \\
\dfrac{2x}{2} = & \dfrac{4}{2} \quad \text{divide} \\
\end{array}
$$

cancel $\quad \dfrac{\cancel{2}x}{\cancel{2}} = \dfrac{\cancel{4}\,2}{\cancel{2}}$

$$
x = 2 \quad \underline{\text{solution}}
$$

10. To check the answer, "x = 2", substitute the value 2 for x.

$$2x + 3 = 7$$
$$2(2) + 3 = 7$$
$$4 + 3 = 7$$
$$7 = 7 \quad \underline{check} \quad \text{Therefore } x = 2 \text{ is a correct solution.}$$

- - - - - - - - - - - -

11. To solve the following equation subtract 3 from both sides. Multiply both sides by 2 and cancel. The answer will appear: "x = 8."

Solve:
$$\frac{x}{2} + 3 = 7$$
$$\underline{\quad - 3 = -3} \quad \text{subtract}$$
$$\frac{x}{2} + 0 = 4$$
$$\frac{x}{2} = 4$$
$$\frac{x}{2} \cdot 2 = 4 \cdot 2 \quad \text{multiply}$$

cancel $\quad \dfrac{x}{\cancel{2}} \cdot \cancel{2} = 8$

$$x = 8 \quad \underline{solution}$$

- - - - - - - - - - - -

12. To check the answer "x = 8" in the above equation, substitute 8 for "x." In the last line 7 = 7, therefore 8 is the correct solution.

$$\frac{x}{2} + 3 = 7$$
$$\frac{(8)}{2} + 3 = 7$$
$$4 + 3 = 7$$
$$7 = 7 \quad \underline{check} \quad \text{Therefore } x = 8 \text{ is a correct solution.}$$

13. Here is a summary stated in two important rules used to solve equations in the previous frames.

(1) You can subtract or add the same number to both sides of an equation and the new equation is <u>equivalent</u> to the original equation.

(2) You can divide or multiply both sides of an equation by the same number, except for zero, and the new equation is equivalent to the original equation.

- - - - - - - - - - - - -

14. Compare the following two equations! How are they different?

$$\text{COMPARE:} \quad x + 3 = 7$$
$$x + 3y = 7$$

In the second equation there is a letter "y" next to the number 3. In the first equation this second letter "y" is missing.

- - - - - - - - - - - - -

15. $x + 3 = 7$ is a first degree equation in <u>one</u> variable.
$x + 3y = 7$ is a first degree equation in <u>two</u> variables.

We will study first degree equations in two variables.
By "first" degree we mean the exponents of the variables are "one."

- - - - - - - - - - - - -

16. The equation in two variables $x + 3y = 7$ can be written as: $x + 3y - 7 = 0$. In the below case we subtract a 7 from both sides of the equation to obtain this version of the equation. It is common to express equations in two variables with a zero term on the right side.

$$
\begin{array}{rcl}
x + 3y & = & 7 \\
-7 & = & -7 \\
\hline
x + 3y - 7 & = & 0
\end{array}
$$
This is equivalent to the original equation.

EXERCISE 5: Linear Equations and Graphs

1. Matching: Place the correct letter from the left column on the corresponding blank line of the right equation.

   (a) first degree equation in one variable $x + 2y = 13$ ____

   (b) first degree equation in two variables $x - 7 = 12$ ____

2. Solve each equation by finding the value of each variable.

   (a) $x + 2 = 9$

   (b) $y - 10 = 8$

   (c) $x/4 + 7 = 10$

3. Rewrite the following equations with a zero on the right side of the equals sign.

   (a) $x + 5y = 8$

   (b) $2x - 4y = -12$

4. True or False:

   (a) You may subtract or add the same number to both sides of an equation and the new equation is equivalent to the original equation.

   (b) You may divide both sides of an equation by zero.

ANSWERS FOR EXERCISE 5:

1. $x + 2y = 13$ (b) first degree equation in two variables

   $x - 7 = 12$ (a) first degree equation in one variable

2. (a) $x = 7$

   (b) $y = 18$

   (c) $x = 12$

3. (a) $x + 5y - 8 = 0$

   (b) $2x - 4y + 12 = 0$

   Note: If we mulitply each side of this equation by ½ then it can be rewritten as: $x - 2y + 6 = 0$

4. (a) True

   (b) False

17. A general form of a first degree equation in two
variables is: $ax + by + c = 0$
Where a, b and c are numbers and a and b are not both
zero. While this general form is commonly used, there is
another way to write a first degree equation in two
variables. We will apply rules we learned in the
next frame.

- - - - - - - - - - - -

18. In this frame we take the equation $y - 5x - 2 = 0$
and solve for "y" using basic rules. First we add 5x
to both sides and then we add 2 to both sides. The final
result is the equivalent equation $y = 5x + 2$

$$
\begin{array}{r}
y - 5x - 2 = 0 \\
\underline{+\ 5x = +5x} \\
y + 0 \ - 2 = 5x \\
\underline{ + 2 = + 2} \\
y + 0 = 5x + 2 \\
y = 5x + 2
\end{array}
$$

- - - - - - - - - - - -

19. The most common form of a first degree equation in two
variables is: $y = ax + b$ where a and b are numbers
and "a" does not equal zero.

$y = ax + b$ is very useful in finding solutions to first degree
equations in two variables. Unlike solutions we found
earlier, the solutions for equations in two variables are not
single numbers but pairs of numbers.

20. Let's consider the equation $y = 3x + 2$.
    What are the values for y when $x = 1$ and when $x = 2$?
    The solutions are derived by substituting the values for x
    in the equation and putting the solutions under y in the
    table.

$$y = 3x \quad + 2 \qquad\qquad y = 3x \quad + 2$$
$$y = 3(1) + 2 \qquad\qquad y = 3(2) + 2$$
$$y = \quad 3 \quad + 2 \qquad\qquad y = \quad 6 \quad + 2$$
$$y = \quad 5 \qquad\qquad\qquad y = \quad 8$$

x	y
1	5
2	8

table

- - - - - - - - - - - - -

21. The solutions in the above problem were: when $x = 1$
    then $y = 5$; when $x = 2$ then $y = 8$.
    What are the values for y when $x = 0$ and when $x = -1$?

$$y = 3x \quad + 2$$
$$y = 3(0) + 2$$
$$y = \quad 0 \;\;+ 2$$
$$y = \quad 2$$

x	y
1	5
2	8
0	2
-1	?

SOLUTIONS: When $x = 0$ then $y = 2$ and when $x = -1$ then
$y = -1$ because $3(-1) + 2 = -3 + 2 = -1$

What is the value for y when x is one-third? The answer
or solution is given in the next frame.

22. We subsitute one-third for x and solve for y.

$y = 3x \quad + 2$

$y = 3(\dfrac{1}{3}) + 2$

$y = \quad 1 \quad + 2$

$y = \quad 3$

x	y
1	5
2	8
0	2
-1	-1
$\dfrac{1}{3}$	3

This table represents only a partial list of <u>ordered pairs</u> in what we call the <u>solution set</u>. There is an infinite number of ordered pairs in the solution set.

- - - - - - - - - - - - - -

23. The ordered pairs in the above table can be plotted as a graph consisting of a set of points. We will graph these points in frame 26. Here is a list of the points in set notation:

$\{ (1, 5), (2, 8), (0, 2), (-1, -1), (1/3, 3) \}$

A <u>rectangular coordinate system</u> (shown on the next page) is used to graph any set of ordered pairs of numbers. A graph consists of a plane divided into four regions called <u>quadrants,</u> which are labelled in Roman numerals. The horizontal number line is called the <u>x-axis</u> because it represents the variable x. The vertical number line is called the <u>y-axis</u> because it represents the variable y. The point where the two lines intersect is called the <u>origin</u> and corresponds to the point (0, 0) because x and y are both zero at the origin.

The x-axis is positive to the right of the origin and negative to the left of the origin. The y-axis is positive above the origin and negative below the origin.

Look at point P in quadrant Roman numeral one. This point P corresponds to a pair of numbers (x,y) where x = 2 and y = 1. Point P (x, y) equals the ordered pair (2, 1). The value 2 is selected for x because point P is located two units to the right of the origin. The value 1 is selected for y because point P is located one unit above the origin. In the fourth quadrant point Q corresponds to the ordered pair of coordinates (1, -3). Can you explain the reason based on the explanation we gave for point P?

24. Rectangular Coordinate System

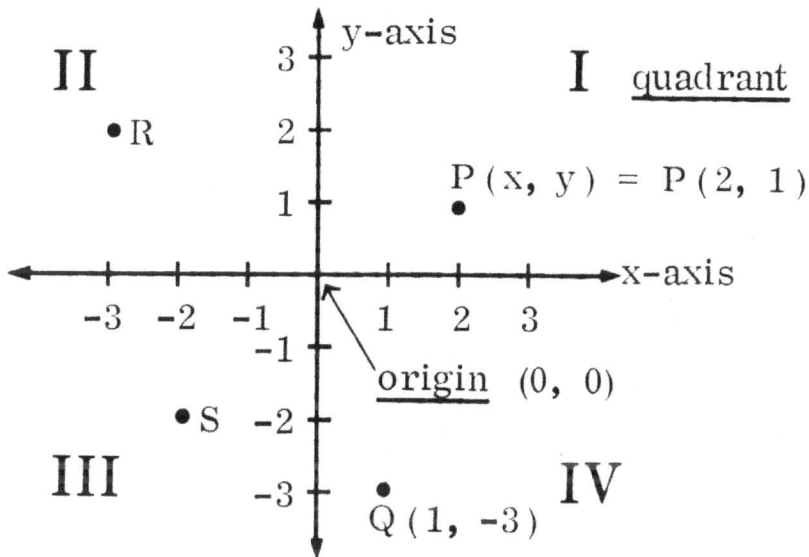

II   I quadrant

• R

P (x, y) = P (2, 1)

-3 -2 -1   1   2   3   x-axis

origin (0, 0)

• S

III   IV

Q (1, -3)

QUESTION: Can you locate the points that correspond to (-3, 2) and (-2, -2)?

— — — — — — — — — — — — — —

25. ANSWERS: (-3, 2) is point R; (-2,-2) is point S. Q corresponds to (1, -3) because point Q is located one unit to the right of the origin directly beneath positive one on the x-axis and point Q is opposite negative three on the y-axis which is three units beneath the origin.

EXERCISE 6: Linear Equations and Graphs

1. Write the two general forms of a first degree equation in two variables

2. Solve each equation for y:

   (a) $y + 2x = 10$

   (b) $y - 7x = 6$

   (c) $y/2 + 3x = 5$

3. If $y = 3x + 2$, fill in the values for y in each of the following tables:

   (a)

x	y
1	
2	
3	

   (b)

x	y
-1	
0	
1/3	

   (c)

x	y
-1/3	
2/3	
5	

4. Locate the following points on the below rectangular coordinate system. You may use your own graph paper.
   $P = (2, -1)$ $Q = (3, 1)$ $R = (-3, -2)$ $S = (-1, 1))$

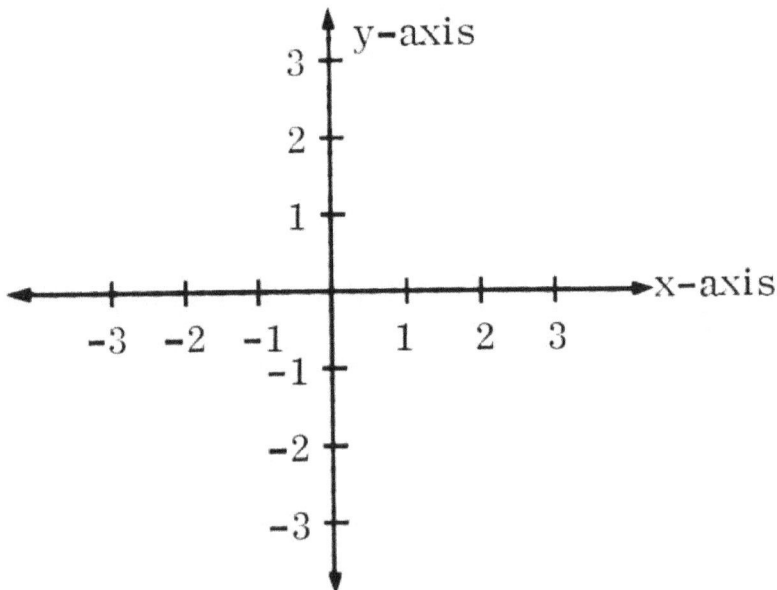

ANSWERS FOR EXERCISE 6:

1. $ax + by + c = 0$ and $y = ax + b$

   Note: In first degree equations the exponent of the variables is one.

2. (a) $y = -2x + 10$

   (b) $y = 7x + 6$

   (c) $y = -6x + 10$

3. (a)

x	y
1	5
2	8
3	11

(b)

x	y
-1	-1
0	2
1/3	3

(c)

x	y
-1/3	1
2/3	4
5	17

4.

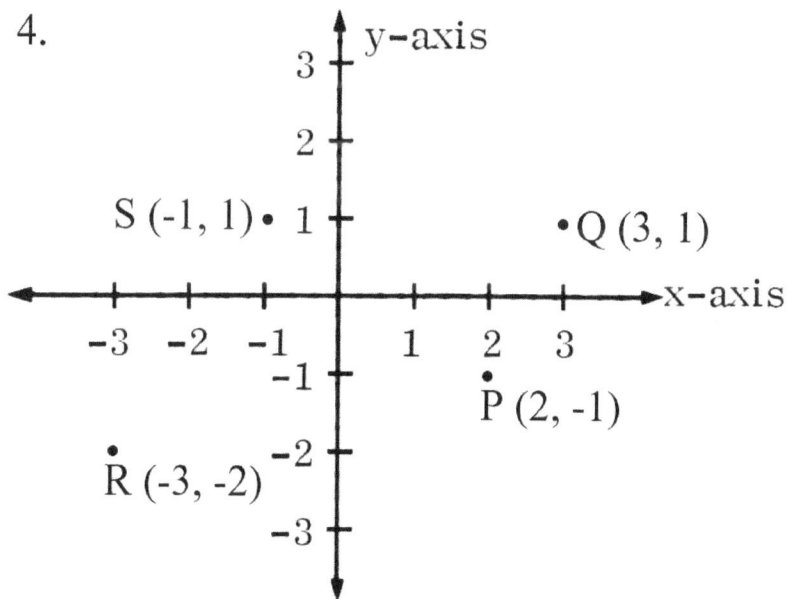

26. The graph of the equation $y = 3x + 2$ is a straight line. This straight line can be seen in the graph at the bottom of this page. We worked with this same equation in previous frames.

First degree equations are called <u>linear</u> equations because their graphs form a straight line. We first construct a table of solutions which is shown below on the left.

We find the values for y by substituting values for x in the equation. Next we plot the ordered pairs in the table as points on thre rectangular coordinate system. For example, the ordered pair (1, 5) is the first entry in the table and it is located at (1, 5) in our below rectangular coordinate system.

When we plot all five points from the table, we find that all of the points are collinear. They lie in a straight line. Therefore, we draw a straight line through all five points. In fact, the graph of $y = 3x + 2$ consists of an infinite set of points located on this straight line. All first degree equations are called linear because their graphs form straight lines.

x	y
1	5
2	8
$\frac{1}{3}$	3
0	2
−1	−1

Graph of
$y = 3x + 2$

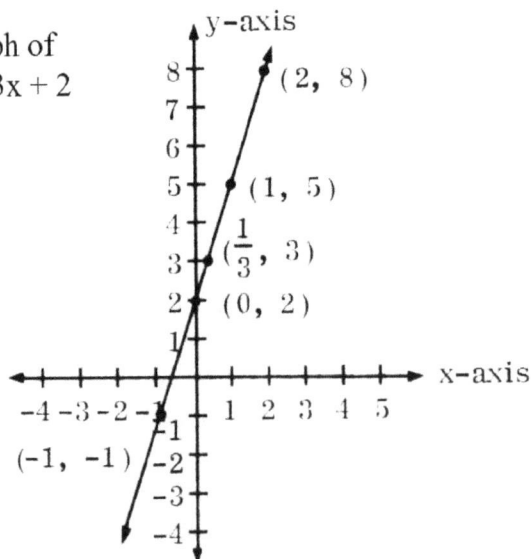

EXERCISE 7: Linear Equations and Graphs

1. Graph $y = 2x - 1$ using the given values for x in the below table. You only need two points to determine a straight line. However, a third point is helpful as a check point to satisfy the property that the points are collinear.

x	y
-1	
0	
+1	

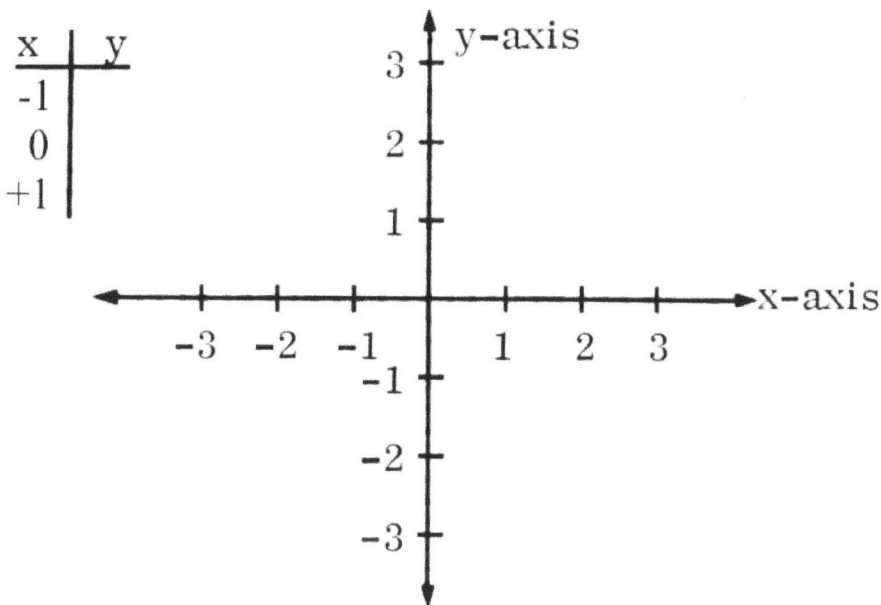

2. Graph each equation on your own graph paper. For each given value of x, find the value of y and plot each point. Draw a straight line through the three points that will represent the equation.

(a) $y = -2x + 1$ for x values: -1, 0, 1

(b) $y = 3x - 3$ for x values: 2, 1, 0

(c) $y = \frac{1}{2}x - 2$ for x values: 3, 2, -2

3. Based on the graphs of the above equations, what effect on the slope of a line does the coefficient of "x" have when you graph a linear equation?

ANSWERS FOR EXERCISE 7:

1.

x	y
-1	-3
0	-1
+1	+1

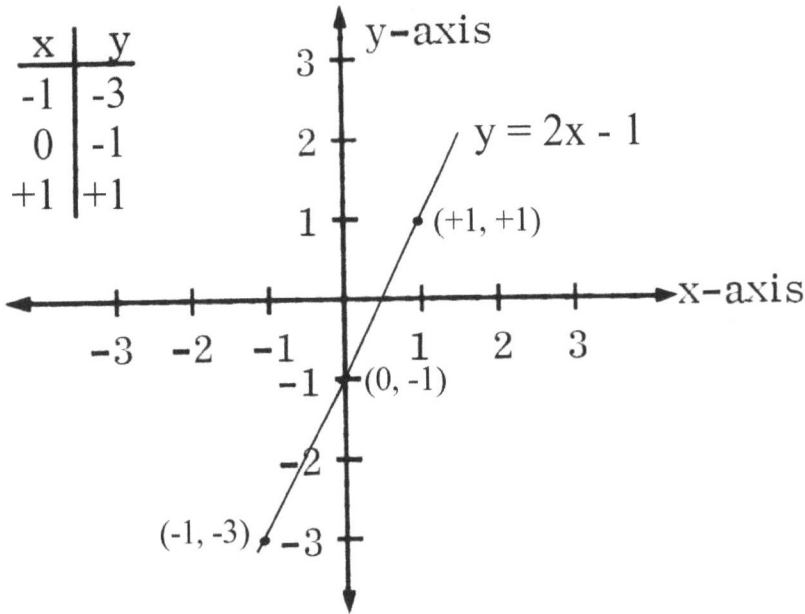

$y = 2x - 1$

(+1, +1)

(0, -1)

(-1, -3)

2. (a)

x	y
-1	+3
0	+1
+1	-1

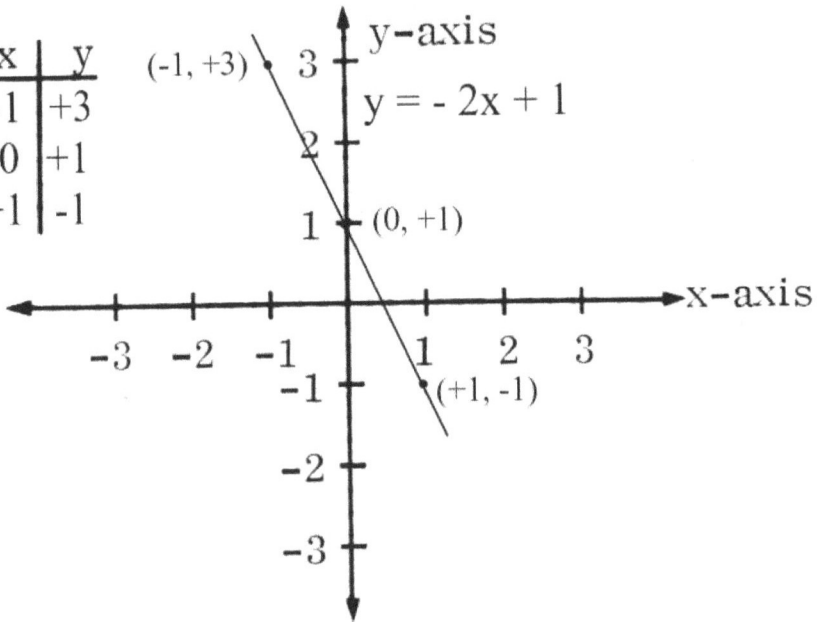

(-1, +3)

$y = -2x + 1$

(0, +1)

(+1, -1)

2. (b)

x	y
2	3
1	0
0	-3

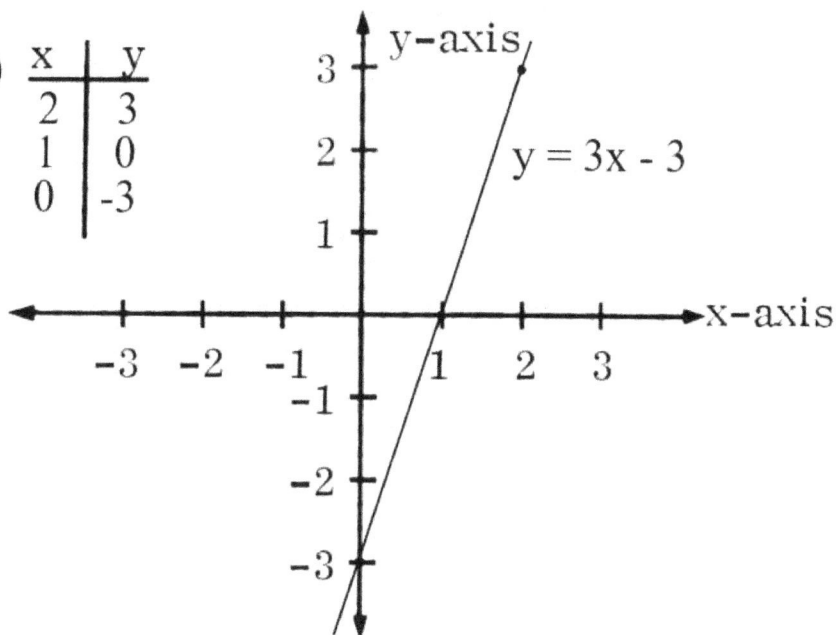

$y = 3x - 3$

2. (c)

x	y
3	$-\frac{1}{2}$
2	-1
-2	-3

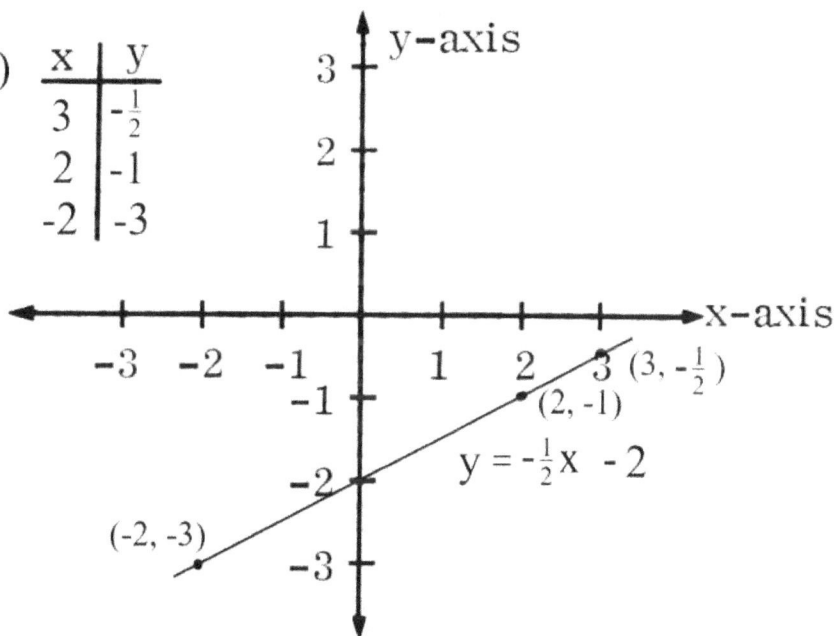

$(3, -\frac{1}{2})$

$(2, -1)$

$y = -\frac{1}{2}x - 2$

$(-2, -3)$

3. If the coefficient of x is positive, then as "x" increases in value, "y" increases in value. If the coefficient of "x" is negative then as "x" increases in value, "y" decreases in value. Compare the slopes of the equations in 2(a) and 2(b) to see the difference.

This concludes *Beginning Algebra.*

For additional books published by

William R. Parks

Visit www.wrparks.com

www.ingramcontent.com/pod-product-compliance
Lightning Source LLC
Chambersburg PA
CBHW081305040426
42452CB00014B/2660